essentials

Springer essentials

Springer essentials provide up-to-date knowledge in a concentrated form. They aim to deliver the essence of what counts as "state-of-the-art" in the current academic discussion or in practice. With their quick, uncomplicated and comprehensible information, essentials provide:

- an introduction to a current issue within your field of expertis
- an introduction to a new topic of interest
- an insight, in order to be able to join in the discussion on a particular topic

Available in electronic and printed format, the books present expert knowledge from Springer specialist authors in a compact form. They are particularly suitable for use as eBooks on tablet PCs, eBook readers and smartphones. *Springer essentials* form modules of knowledge from the areas economics, social sciences and humanities, technology and natural sciences, as well as from medicine, psychology and health professions, written by renowned Springer-authors across many disciplines.

More information about this subseries at http://www.springer.com/series/16761

Torsten Schmiermund

Introduction to Stereochemistry

For Students and Trainees

Springer

Torsten Schmiermund
Frankfurt am Main, Germany

ISSN 2197-6708 ISSN 2197-6716 (electronic)
essentials
ISSN 2731-3107 ISSN 2731-3115 (electronic)
Springer essentials
ISBN 978-3-658-32034-8 ISBN 978-3-658-32035-5 (eBook)
https://doi.org/10.1007/978-3-658-32035-5

Responsible Editor: Désirée Claus
This Springer imprint is published by the registered company Springer Fachmedien Wiesbaden GmbH part of Springer Nature.
The registered company address is: Abraham-Lincoln-Str. 46, 65189 Wiesbaden, Germany

What You Can Find in This *essential*

- A short introduction to the stereochemistry of organic compounds
- Explanation of most of the intentions that make stereochemical statements
- Differences between configuration, conformation, constitution isomerism and tautomerism
- Distinction of geometric isomerism and optical isomerism
- Different display/visualization options for isomers

Contents

Stereochemistry—Why?

1

As early as 1808, a tetrahedral arrangement for molecules of the type AB_4 was postulated (W. H. Wollaston), optical rotation was discovered in 1813 (J.-B. Biot) and the first racemic fission was performed in 1848 (L. Pasteur). Le Bel and van't Hoff developed the actual theory of the spatial arrangement of atoms in molecules in 1874.

Stereochemistry ("space chemistry," "spatial chemistry") is a branch of chemistry that deals with the three-dimensional structure of molecules. Basically, two aspects are distinguished:

The three-dimensional structure of molecules that have the same composition (molecular formula) but a different arrangement of atoms is called stereochemical isomerism or static stereochemistry.

The spatial course of chemical reactions, especially those involving stereoisomeric molecules, is called stereochemical dynamics or dynamic stereochemistry.

Until the middle of the twentieth century, the main focus was on static stereochemistry. With improved analytical methods, the relationships between spatial structure, reactivity and reaction mechanism began to play an increasingly important role. Today, stereospecific and stereoselective reactions and syntheses are indispensable in the search for new drugs and plant protection products. Similarly, many biological processes can only be adequately understood by the three-dimensional structure of the substances involved.

This book only deals with basic static stereochemistry. For additional details, please refer to the bibliography and corresponding special literature.

© Springer Fachmedien Wiesbaden GmbH, part of Springer Nature 2021
T. Schmiermund, *Introduction to Stereochemistry*, Springer essentials,
https://doi.org/10.1007/978-3-658-32035-5_1

1.1   Isomerism—What is it?

Isomers (Greek *isos* = equal, *meros* = part, "from equal parts") are chemical compounds with the same elementary composition but different structure. This isomerism can consist in a different linkage of the atoms among themselves, i.e., in a different order. This is called constitutional or structural isomerism.

But isomerism is also possible with the same elemental composition and the same arrangement (same atomic sequence). One speaks here of stereoisomerism. Here the atoms of the isomers are spatially differently arranged.

1.2   Auxiliary Means

First you have to learn to translate the different ways of representing stereochemistry into a picture, which—although drawn two-dimensional—is still imaginable three-dimensionally in your own imagination. This requires time, practice, imagination and also a bit of hard work.

Classic, "typical" aids are usually molecular modelling sets. If you are lucky enough to be able to use them or to afford them: go ahead. But preferably use so-called *ball-and-stick* models. Rigid spherical models are less suitable. Everyone else can easily make their own molecular modelling sets by themselves.

For this, you need:

- For the atoms, you can use for example: halved corks from wine bottles, polystyrene balls, colored modelling clay (use a non-drying modelling clay) or whatever else you think is suitable and have at hand.
- If you want to use wooden or plastic balls, it is recommended to pre-drill the appropriate holes.
- For the bindings are an ideal fit: Toothpicks, pieces of wire (wire thickness minimum 0.5 mm, maximum 2 mm), thin plastic straws (cut to fit), or similar

If you have procured your material, the atoms must still be color-coded. Paint your corks, balls, etc. as followed:

- Carbon (C): black (24)
- Hydrogen (H): white (30)
- Oxygen (O): red (12)
- Nitrogen (N): blue (4)
- Fluorine (F): light green (1)

- Chlorine (Cl): green (4)
- Bromine (Br): orange (2)
- Iodine (I): purple (2)

The number in brackets indicates the minimum number of atoms of the respective species you should prepare. If you have prepared too few atoms, you may have to disassemble the models of molecules you want to look at in order to make other molecules.

For our purposes, you can use all atoms of the same size. A diameter of 2–3 cm is conveniently sufficient. The bonds should be about twice to three times the length. So the bond reaches about to the center of the atom and an atomic diameter of a "visible bond" remains.

The different materials have advantages and disadvantages. Modelling clay, for example, is relatively heavy and soft at the same time. Larger molecules only stick together poorly. In return, you do not have to paint anything. Corks are sometimes very strong and therefore difficult to join, polystyrene can break out relatively easily. You may have to try different variations. Models made of cardboard or paper may also prove to be helpful. The rule is: you have to get along with the models. Simple availability takes precedence over perfect appearance.

1.3   How do I Construct a Molecule?

The following sequence has proved to be successful:

1. Assemble the carbon skeleton (including double bonds).
2. Prepare all functional groups separately.
3. Dock hetero atoms with double bonds to the C skeleton.
4. Add the functional groups.
5. Finally, add the hydrogen atoms.

Please remember that the carbon atoms in particular may have different bond angles depending on the bond condition/hybridization.

Basics

2

On the following pages, we will be confronted with a certain number of different models or representations. Each of these representations has arisen from different problems.

To get along with the different two-dimensional representations and the corresponding construction of three-dimensional molecular models, first of all some basic things. An overview of the different forms of isomerism can be found in Fig. A.1 at the end of the book. In Fig. A.2, you will find a table showing the differences and similarities of the isomers.

2.1   What is Assumed

Assumed to be known:

- Valence/binding nature of the individual elements (especially C, H, O, N, S, P, halogens)
- Oxidation numbers and atomic numbers
- Hybridization on the C-atom (sp^3-, sp^2- and sp-hybrid orbitals and their formation)
- Binding angles and distances
- Binding types (σ- and π bindings)
- Different formula notations (e.g., with/without H atoms)
- Basic knowledge of the nomenclature of organic compounds

© Springer Fachmedien Wiesbaden GmbH, part of Springer Nature 2021
T. Schmiermund, *Introduction to Stereochemistry*, Springer essentials,
https://doi.org/10.1007/978-3-658-32035-5_2

2.2   Terms of Access

Let us first deal with some important terms before we look at the different possibilities of the spatial structure of organic molecules in detail.

2.2.1   Isomer

Isomers are substances of the same composition (same molecular formula) but with different atomic arrangements or different spatial molecular structures.

2.2.2   Constitution and Constitutional Isomers

The **constitution** (lat. *constituere* = to set up) indicates the type of bonds (single, double, triple) and the mutual bonding of the atoms in the molecule. The molecular formula always remains the same. Possible differences in the spatial arrangement (such as rotations around a single bond) are not taken into account for constituent isomers.

Constitutional isomers therefore have a different linkage between the atoms for the same molecular formula. They have a different constitution.

Constitutional isomers include:

- Skeletal isomers (different structure of the carbon skeleton)
- Position isomers (different position on the carbon skeleton)
- Functional isomers (different functional groups)
- Valence isomers (different bonds)
- Tautomers (different bonds and different functional groups)

2.2.3   Conformation and Conformational Isomers

The **conformation** (lat. *conformare* = to shape, to form accordingly; corresponds approximately to the structural formula) represents the spatial arrangement of all atoms of a molecule with a defined configuration, which can be created by rotation around single bonds, but cannot be brought into alignment.

Conformation isomers can only be isolated at room temperature if the energy threshold between the conformation isomers exceeds 70–85 kJ mol^{-1}.

2.2.4 Configuration and Stereoisomers

The **configuration** (lat. *configurare* = to form uniformly) indicates the spatial arrangement of the atoms of a molecule. Forms that are created by rotating atoms around single bonds are *not* taken into account.

Stereoisomers (Greek *stereo* = rigid; in the sense of: spatial, three-dimensional), also called configuration isomers, have the same molecular formula and the same constitution. They differ in the spatial arrangement of their atoms and therefore have a different configuration.

2.2.5 Tautomerism

Tautomerism (from Greek *tauto* = the same and *meros* = proportion) is the rapid reversible transition from one constitutional isomeric form to another.

Important tautomerisms are:

- Keto-enol tautomerism (for aldehydes/ketones)
- Amide-imidol tautomerism (for amides)
- Oxo-enol-enon tautomerism (for heteroaromatics)
- Valence tautomerism (for polyenes)

Constitutional Isomers

3

3.1 Skeletal Isomerism

In skeletal isomers, the isomers only differ in the arrangement of the carbon atoms relative to each other. They are also called skeletal isomers because the carbon skeleton of the respective compound is changed.

From the molecular formula C_6H_{14}, five isomeric hexanes, which can be distinguished in their physical properties, can be represented (Fig. 3.1).

3.1.1 *n-, iso-, neo-*

In the older nomenclature, the arrangement of chain-shaped hydrocarbons was marked with the prefixes *n-, iso-* and *neo-*. These prefixes are no longer used in the systematic nomenclature. However, they are still frequently used in laboratory language and as a part of trivial names (see also Fig. 3.1).

Unbranched hydrocarbons are called *n*-compounds (n for "normal"). The prefix *iso-* (Greek: *isos* = equal) denotes a branched C framework without additional details and was only used for the *iso*-propyl radical (*i*-Pr-). The prefix *neo-* (Greek *neos* = new) was used as a—likewise unspecific—prefix for "new," mostly synthetically produced compounds. According to IUPAC, it was only approved for the neopentyl radical or neopentane (2,2'-dimethyl-propane).

© Springer Fachmedien Wiesbaden GmbH, part of Springer Nature 2021
T. Schmiermund, *Introduction to Stereochemistry,* Springer essentials,
https://doi.org/10.1007/978-3-658-32035-5_3

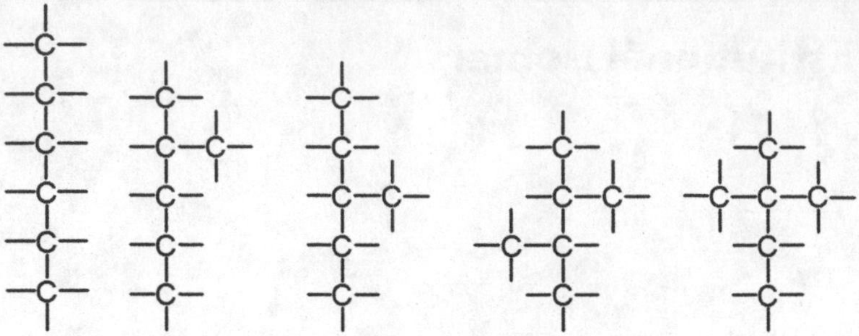

| Hexane | 2-Methyl-pentane | 3-Methyl- | 2,3-Dimethyl-butane | 2,2'-Dimethyl butane |
| (*n*-hexane) | (iso-hexane) | pentane | ("Di-iso-propyl") | (neo-hexane) |

Fig. 3.1 Skeletal isomers hexanes

3.2   Positional Isomerism

In positional isomers, functional groups are located at different positions in the carbon skeleton. The carbon skeleton itself and the properties of the functional groups do not differ. In this form of isomers, the isomers are easily distinguishable from each other too due to their physical properties.

In place of the historical prefixes presented here, the numerical positional data according to IUPAC should be preferred today.

3.2.1 *gem.-, vic.-* Isomers

Geminal (abbreviation: *gem.*; lat. *gemini* = twin) means that two similar substituents are fixed to the same C-atom. Vicinal (abbreviation: *vic.;* lat. *vicinus* = neighbor) means that the two similar substituents are located on two neighboring C-atoms. If there is at least one C-atom between the C-atoms with similar substituents, these are isolated substituents. There is no existing separate historical designation for these (see Fig. 3.2).

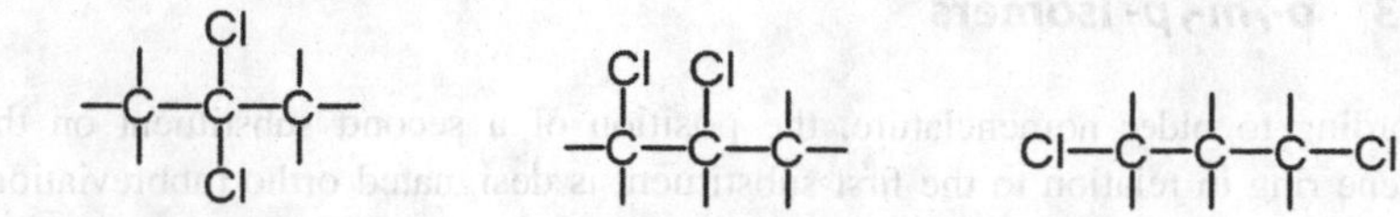

2.2-Dichloropropane
(*gem.* Dichlorpropane)

1.2-dichloropropane
(vic. dichlorpropane)

1.3-dichloropropane

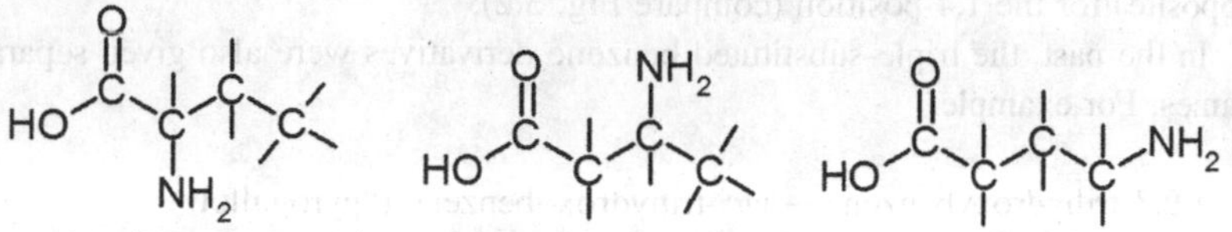

2-amino-butanoic acid
(α-amino butyric acid)

3-amino-butanoic acid
(ß-amino butyric acid)

4-amino-butanoic acid
(γ-amino butyric acid)

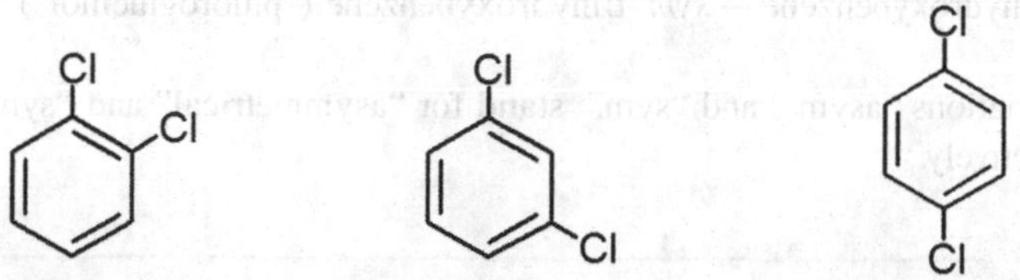

1.2-dichlorobenzene
(o-dichlorobenzene)

1.3-dichlorobenzene
(m-dichlorobenzene)

1.4-dichlorobenzene
(p-dichlorobenzene)

Fig. 3.2 Position isomers

3.2.2 α-, β-, γ-, ω- Isomers

If a further functional group is added to a parent compound that already contains a functional group (e.g., an acid or an alcohol), the position of the new/second substituent is marked with Greek lower case letters (compare Fig. 3.2). It then means:

- α-: Second substituent on the directly adjacent C-atom
- β-: Second substituent on the next but one C-atom
- γ-: Second substituent on the third C-atom
- ω-: Second substituent at the most distant possible C-atom

3.2.3   *o-, m-, p-* Isomers

According to older nomenclature, the position of a second substituent on the benzene ring in relation to the first substituent is designated ortho (abbreviation: *o-*; Greek *ortho* = straight) for the 1,2-position, meta (abbreviation: *m-*; Greek *meta* = between) for the 1,3-position and para (abbreviation: *p-*; Greek *para* = opposite) for the 1,4-position (compare Fig. 3.2).

In the past, the triple-substituted benzene derivatives were also given separate names. For example:

- 1,2,3-trihydroxybenzene → *vic.*-trihydroxybenzene ("pyrogallol")
- 1,2,4-trihydroxybenzene → *asym.*-trihydroxybenzene   ("hydroxyhydroquinone")
- 1,3,5-trihydroxybenzene → *sym.*-trihydroxybenzene ("phloroglucinol")

The abbreviations "asym." and "sym." stand for "asymmetrical" and "symmetrical," respectively.

3.3   Functional Isomerism

If different functional groups can be represented by the same empirical formula, it is called functional isomers. Consequently, these isomers belong to different classes of substances with different chemical and physical properties. Functional isomers are "isomers by definition." Normally compounds with the same empirical formula but different functional groups are not regarded as isomers because they belong to different classes of substances.

Examples:

- C_2H_6O: ethanol (CH_3CH_2–OH)//dimethyl ether (H_3C–O–CH_3)
- C_3H_6O: acetone (H_3C–C(O)–CH_3)//acetaldehyde (H_3C–CH_2–CHO)
- $C_4H_8O_2$: ethyl acetate (H_3C–C(O)–O–C_2H_5)//4-hydroxy-butan-2-one (H_3C–C(O)–C_2H_4–OH)

3.4   Valence Isomerism

In the valence or bond isomers, the number and/or position of σ and π bonds differ. To convert the isomers into each other, bonds have to be opened and rebonded.

Sometimes skeleton isomers occurs at the same time as valence isomers (see also Sect. 3.5.4 Valence tautomerism).

$$R-\underset{\underset{H}{|}}{\overset{\overset{R'}{|}}{C}}-\overset{5}{CH}=\overset{4}{CH}-\overset{3}{CH}=\overset{2}{CH}-\overset{1}{CH_3} \rightleftharpoons \underset{R}{\overset{R'}{\diagdown}}C=\overset{5}{CH}-\overset{4}{CH}=\overset{3}{CH}-\overset{2}{CH_2}-\overset{1}{CH_3}$$

Other examples are:

- C_3H_4: propadiene ($H_2C=C=CH_2$)//propine ($HC\equiv C-CH_3$)//cyclopropene
- C_6H_8: cyclohexa-1,2-diene//cyclohexa-1,3-diene//cyclohexa-1,4-diene

3.5 Tautomerism

Tautomerism is, so to speak, simultaneously a special case of valence isomers *and* functional isomers. Tautomers are constitutional isomers that are quickly and reversibly transformed into each other by the migration of individual atoms or groups of atoms. The tautomeric forms are in a dynamic equilibrium. This leads to a constant ratio of the tautomers to each other. Therefore, these isomers usually cannot be separated from each other. The differentiation of the tautomeric forms is usually done by spectroscopic methods.

If a hydrogen atom (proton) changes its place within the molecule, it is also called prototropy or proton isomerism.

3.5.1 Keto-enol Tautomerism

Keto-enol tautomerism is the most common form of tautomerism. It occurs when a proton is split off from the α C-atom of a carbonyl group and the carbonyl oxygen is then protonated.

In most cases, the balance is on the side of the keto form. For example, propanone (acetone) is present in the enol form at about 3 ppm, ethyl 3-oxybutyrate (acetoacetic ester) at 8%.

Keto-Form   **Enol-Form**

For some substances, however, the enol form predominates. For example, pentane-2,4-dione (acetylacetone), 85% of which is in the enol form (4-hydroxy-pentan-2-one). This is explained by the formation of an intramolecular hydrogen bond due to the stabilization.

2,4-Pentandione   **4-Hydroxy-pentan-2-one**
Keto-Form   **Enol-Form**
15%   **85%**

A special form of keto-enol tautomerism, the ketol-endiol tautomerism, occurs with α-hydroxy ketones (acyloins). Thus, hydroxypropanal turns into propendiol, which in turn can isomerize into hydroxypropanone.

α-Hydroxy-aldehyd   **Endiol**   **α-Hydroxy-keton**
(2-Hydroxy-propanal)   **(1,2-Propendiol)** **(1-Hydroxy-propanon)**

The 2-hydroxy-propanal and 1-hydroxy-propanone only differ in the position of the carbonyl oxygen and the hydroxy group and are therefore positional isomers. However, all three compounds are also functional isomers of each other (aldehyde, alcohol and ketone).

3.5.2 Amide-imidol Tautomerism

The tautomers of amides are called imidols (also: imidic acids). Here, one proton of the amine function migrates to carbonyl oxygen.

Amid   Imidol

3.5.3 Oxo-enol-enon Tautomerism

Spectroscopic methods can also be used to determine tautomeric equilibria in heteroaromatics. Oxo-enol-enon tautomerism is best known for pyrazolone-(5) derivatives (2-oxo-4,5-dihydropyrazoles), which are important intermediates for drugs.

Oxo   Enol   Enon
(CH-Form)   (OH-Form)   (NH-Form)

3.5.4 Valence Tautomerism

The concept of valence tautomerism (also called binding tautomerism) is usually not sharply separated from the concept of valence isomers: Both notions are often used synonymously. If the isomerization leads to a product that cannot be distinguished from the starting compound, the isomers are often called degenerated valence isomers. Compounds whose bonds are reversibly transformed into each other ("fluctuating") and thereby form degenerate valence isomers are here - for better differentiation - as valence tautomers refered to as.

A typical example is the nonaromatic 1,3,5,7-cyloocta-tetraene. In addition to the valence tautomerism (or degenerate valence isomers), valence and framework isomers to *cis*-bicyclo-[4,2,0]octatrien-(2,4,7) (present at equilibrium at about 100 ppm) can be observed.

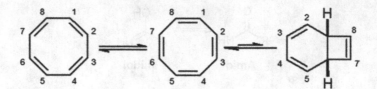

Stereoisomerism

4

Stereoisomers not only have the same molecular formula but also the same structure (constitution). The isomers differ in the spatial arrangement (configuration) of their atoms.

Stereoisomers can be divided into:

- Configuration isomers: geometric isomers
- Configuration isomers: optical isomers
- Conformation isomers

According to Dale's definition (Dale 1978), configuration isomers are molecules that can only be transformed into each other by splitting and forming new bonds.

Accordingly, conformation isomers merge into each other without splitting bonds. The folding process of a cyclohexane ring system—the so-called ring inversion—can formally be regarded as a simultaneous rotation around all C–C bonds.

Stereoisomers can be enantiomers or diastereomers of each other.

4.1   Enantiomers

Enantiomers (Greek *enantios* = counterpart, *meros* = part) are isomers that behave like image and mirror image in their spatial structure. This type of isomerism is therefore also called chirality (Greek *chiros* = hand, "handedness").

This can clearly be seen on body parts (right/left hand) or objects (right/left shoe, right/left turned screws). The two hands, for example, are not congruent, but behave like image and mirror image to each other.

© Springer Fachmedien Wiesbaden GmbH, part of Springer Nature 2021   17
T. Schmiermund, *Introduction to Stereochemistry,* Springer essentials,
https://doi.org/10.1007/978-3-658-32035-5_4

Enantiomer pairs have the same physical properties (density, solubility, melting and boiling point, etc.). However, they differ in their optical activity and biological effects. Due to the different optical activity of the enantiomers, they are also called optical isomers or optical isomerism. In the (older) German literature, enantiomers are also called antipodes (Greek *anti* = against, *podos* = foot), because they rotate polarized light by the same amount but in different directions.

4.1.1 Optical Activity

Certain materials are able to rotate the plane of linear polarized light. This property is called optical activity. The process can be explained by the fact that the electrons in a molecule cannot oscillate equally in all directions. In different directions, this results in different polarizability of the electrons. The electrons are therefore excited to oscillate differently—depending on the configuration of the optically active substance. This leads to a rotation of the polarized light to the right or left, which can be measured as an angle (angle of rotation, α) with a polarimeter after it has been emitted.

If the plane of the light is turned clockwise, the substance is called dextrorotatory and is marked with (+). If the plane of light is turned counter-clockwise, the substance is therefore levorotatory and (−). The direction of rotation cannot be derived from the absolute configuration, but must be determined for each pair of enantiomers using a polarimeter.

For a substance to have optical activity, it must have at least one stereo center (also called chirality center or stereogenic center). The stereo center does not necessarily have to be a single atom of the molecule, but can also be located between several atoms—as is the case with double bonds, for example.

If the stereogenic center is a carbon atom with four different substituents, e.g., chlorofluoroiodomethane, this carbon atom was previously called an asymmetric carbon atom and marked with an asterisk (C*) in structural formulas. As this assignment is not entirely clear, it should no longer be used.

4.1.2 Racemates, Racemic Mixtures

The term racemate is derived from the Latin name of racemic acid *(acidum racemicum),* as this is the first time a racemate has been separated into its two enantiomers, namely L-(+)-tartaric acid and D-(−)-tartaric acid (so-called racemate cleavage).

Mixtures of equal proportions of the (+) and (−) forms are called racemates and are identified by (±) or the prefix *rac*-. As both isomeric forms have the same optical activity but differ in their sign, the optical activity of the racemate is zero. If both forms are present in a mixture of enantiomers but in a ratio different from 1:1, this is called a non-racemic mixture and sometimes the prefix *ambo*- (lat.: *ambo* = both) is used. In contrast to racemates, non-racemic mixtures show optical activity.

4.1.3 Biological Effect

The different biological effects of optical isomers ("mirror images") can for example be seen in:

- Odor: (S)-(+)-carvone smells like caraway, (R)-(−)-carvone smells like mint
- Taste: (S)-Valin tastes bitter, (R)-Valin tastes sweet
- Pharmacology: (S)-(−)-thalidomide is reprotoxic, (R)-(+)-thalidomide is a sleeping pill with almost no side effects. (Unfortunately, in the human body the isomers transform into each other, because they racemize.)

4.2 Diastereomers

Diastereomers (long form: diastereoisomers) are stereoisomers that are not enantiomers, i.e., they are not enantiomeric to each other. Diastereomers have different physical properties and can thus be separated.

4.2.1 Diastereomers and Geometric Isomerism

Double bonds are rigid due to the π bonds, a rotation cannot take place. Therefore, groups bound to them are also not freely rotatable. This results in different spatial arrangements (*cis*-, *trans*- or (*Z*)-, (*E*)- configurations) and the same is true for monocyclic or unbridged bicyclic compounds.

With bridged bicyclic compounds, *syn*-, *anti*-isomerism occurs in addition to *endo*-, *exo*-isomerism (see Sect. 5.5).

4.2.2 Diastereomers and Several Stereo Centers

Compounds with several stereo centers may differ in each stereo center with respect to their spatial structure. If the configuration of the compounds is different in *all* stereo centers, they are enantiomers. These are not counted as diastereomers.

If the compounds differ in exactly one stereo center, this special form of diastereomerism is also called epimericity (Greek *epi* = after, at, on, near), the two stereoisomers are called epimers.

If the stereo centers in a molecule are similar, i.e., the respective stereo centers have the same functional groups, there is a mirror plane in the molecule itself. This is where so-called *meso* compounds (Greek *meso* = central) exist. In contrast to the other diastereomers, these are usually achiral, i.e., optically inactive.

Figure 4.1 shows an example of the isomers of tartaric acid. Please note that L- tartaric acid and D- tartaric acid are enantiomers, whereas *meso*-tartaric acid is a diastereomer to L- *and* D- tartaric acid.

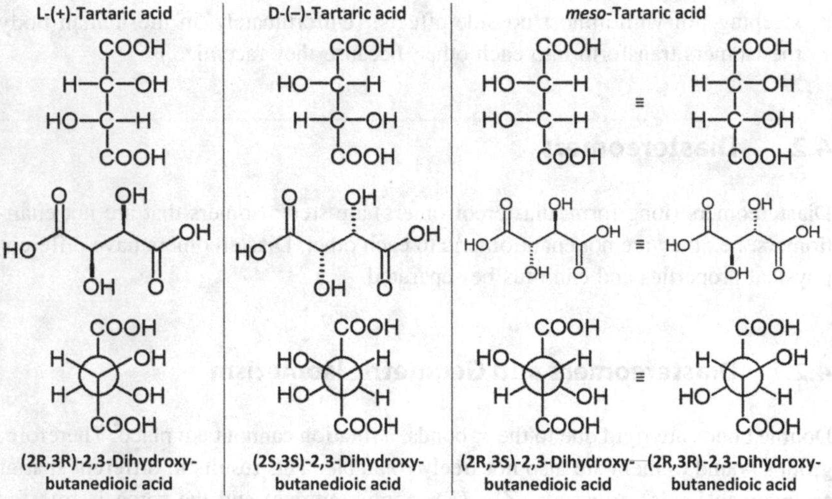

Fig. 4.1 Representations of the diastereomeric tartaric acid forms: Fischer projection, wedge-dash formula and Newman projection. Note: The racemate of tartaric acid, the (±)-tartaric acid, is also called racemic acid. It must not be confused with meso-tartaric acid

4.2.3 Diastereomers and Enantiomers

Diastereomers can be distinguished by their physical properties such as melting point or solubility. Enantiomers only differ in the sign of the angle of rotation.

If a compound contains n stereo centers, 2^n configuration isomers are possible. For aldotetroses, for example, this means $2^2 = 4$, for aldohexoses $2^4 = 16$ isomers (see Fig. 4.2 and A.3). There are thus 2^{n-1} (or $2^n/2$) enantiomeric pairs. Their number is reduced if there is a mirror plane within a molecule due to similar stereo centers and therefore *meso*-forms exist.

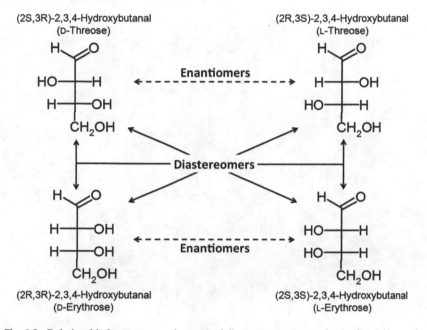

Fig. 4.2 Relationship between enantiomers and diastereomers (example: erythrose/threose)

Configuration Isomers: Geometric Isomerism

5

The geometric isomerism is a special case of configuration isomerism and will be considered separately here. Most geometric isomers have a center of symmetry or a plane of symmetry ('mirror plane'), so that they are achiral compounds. This distinguishes them from the chiral or asymmetrical configuration isomers.

5.1   *cis-/trans-* Isomerism in Double Bonds

The *cis-*, *trans* isomerism is caused by different arrangement of the substituents to each other, with respect to the double bond in the molecule, which serves as a reference plane. The substituents can be arranged on the same side (lat. *cis =* this side) or on opposite sides (lat. *trans =* beyond).

However, the assignment is only clear if a hydrogen atom and a different substituent are bonded to each of the two C-atoms of the double bond. Only then this over the time developed historically positional data can be used. If there are more than two substituents, the (*Z*)-/(*E*)- nomenclature must be used.

cis-1,2-Dichloro-ethene	H H C=C Cl Cl	H Cl C=C Cl H trans-1,2-Dichloro-ethene
cis-Butene-diacid, cis-Ethen-1,2-dicarbonic acid, maleinic acid	H H C=C HOOC COOH	H COOH C=C HOOC H trans-Butene-diacid, trans-Ethen-1,2-dicarbonic acid, fumaric acid

5.2   cis-/trans- Isomerism in Ring Systems

The *cis-*, *trans* isomerism results from the position of the two substituents in relation to the molecular plane. If two substituents are on the same side of the molecular plane, the *cis-* compound is involved; if they are on opposite sides, the *trans-* form is present.

The two diastereomeric forms can be separated from each other, as bonds would have to be broken and reattached for mutual transformation. For example, *cis*-1,2-dimethylcyclopentane has a boiling point of 99 °C and the trans-compound of 92 °C.

trans-   **cis-**

1,2-Dimethylcyclopentan

The recognition of the respective form is somewhat more difficult with substituted cyclohexanes (see Sect. 8.4). If both substituents are either equatorial (e) or axial (a) to the ring plane, then we are dealing with the trans-compound. If the two substituents are arranged differently to the ring plane (i.e., a,e or e,a), then it is the *cis*-compound.

cis-1,4-Dimethyl-cyclohexan (e,a)   trans-1,4-Dimethyl-cyclohexan (e,e)

In condensed bicyclic hydrocarbons, stereoisomerism also occurs as *cis-/trans-*isomerism. This is, for example, the case with decalin (bicyclo[4.4.0]decane or decahydronaphthalene). The *trans*-decalin is rigid and has a boiling point of 185 °C. The *cis*-decalin is flexible and can therefore transform into its enantiomer (its mirror image isomer); it has a boiling point of 194 °C.

trans-Decalin

cis-Decalin

5.3 CIP Rules

The Cahn-Ingold-Prelog Convention (CIP Convention for short) was signed in 1966 by R. S. Cahn, C. K. Ingold and V. Prelog to clearly describe the spatial arrangement of different substituents on atoms or double bonds. A revision took place in 1982 by V. Prelog and G. Helmchen. Due to these rules, a clear naming in (Z)- or (E)-isomers is possible. This is also called the "absolute configuration."

Rules of the CIP Convention

- In principle, all substituents of the individually considered C-atom are sorted according to their atomic number (AN): The larger the AN, the higher the priority.
- One moves outward from the respective C-atom step by step - and thus forms so-called "spheres"—until a clear assignment is possible.
- For compounds containing **stereo centers** ("asymmetric C-atoms") (see Sect. 6.2), the substituents bound to this C-atom form the sphere a. The substituent with the lowest priority is thought to be turned away from the observer ("backward").
- If there is a **double bond** in the molecule, the atoms bound to the C-atoms carrying the double bond are those of sphere a. They are also sorted by falling AN. The atom with the highest atomic number has first/highest priority.
- Multiple bonds are counted as multiple single bonds. For example, $-C=O$ becomes $-C-(O-O)$ and $-C\equiv N$ becomes $-C-(N-N-N)$. That means: If an atom is connected by a double or triple bond, the AN is multiplied by two or three.
- If one does not come to a clear result here yet, one moves one bond level further outward ("atoms of the second sphere"/"sphere b").
- If necessary, one goes even further outward (spheres c, d, …)

Fig. 5.1 Determination of
the order of priority
according to CIP. Shown are
the spheres (a, b and c) as
well as the atomic numbers
(corrected to the number of
bonds) of the respective
atoms

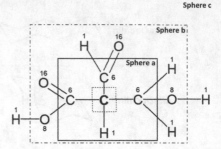

- If different isotopes of an element are contained in the compound, the isotope
  with the higher mass also has higher priority.
- For the most important substituents, the order of priority is decreasing:

  $$I > Br > Cl > SH > OH > NH_2 > COOH > CHO > CH_2OH > CN > CH_2NH_2 > CH_3 > H$$

Example

Explanation of the sequence using the example of 2-formyl-3-hydroxy-propionic
acid ($=HOOC-CH(CHO)-CH_2OH$; structural formula and spheres in Fig. 5.1):

- The C-atoms of the substituents **COOH**, **CHO** and **CH₂OH** initially (**sphere a**)
  have the atomic number 6, the hydrogen has the atomic number 1.
- According to the atomic numbers, in the **sphere b:**
  - For **COOH**: 8×3 (oxygen; 1 single bond, 1 double bond) $= 24$.
  - For **CHO** then 8×2 (oxygen, 1 double bond) $= 16 + 1$ (hydrogen) $= 17$.
  - For **CH₂OH** 1×8 (oxygen) $= 8 + 2$ (hydrogen) $= 10$.
- The missing hydrogen atoms of COOH and CH₂OH belong to **sphere c.** As a
  clear order of priority has already been established, they do not need to be taken
  into account (in this example).
  - The order of priority is therefore: $COOH > CHO > CH_2OH > H$

5.4   (Z)-/(E)-Isomerism

Using the CIP rules, it is now possible to clearly identify compounds where the *cis-/trans-* nomenclature fails. Here, the (Z) stands for "together" (German *zusammen*) and the (E) for "opposite" (German *entgegen*) and always refers to the relative position of the two substituents with the highest priorities.

(Z)-But-2-en, *cis*-2-Buten	H_3C, H \ C=C / H, CH_3	H, H_3C \ C=C / CH_3, H	(E)-But-2-en, *trans*-2-Buten
(Z)-1-Chlor-2-methyl-buten	Cl, H \ C=C / C_2H_5, CH_3	Cl, H \ C=C / CH_3, C_2H_5	(E)-1-Chlor-2-methyl-buten
(Z)-1-Brom-2-chlor-1-iod-ethen	Cl, H \ C=C / I, Br	Cl, H \ C=C / Br, I	(E)-1-Brom-2-chlor-1-iod-ethen

If there are several double bonds in a molecule, (Z)- or (E)- must be indicated for each double bond.

Note:

- Frequently, the (Z)- and *cis-* or (E)- and *trans*-configurations are often identical. But this does not always has to be the case.
- Cumulative double bonds with an *odd* number of double bonds form *cis-/trans-* or (Z)-/(E)-isomers.
- Cumulative double bonds with an *even* number of double bonds form axially chiral molecules and thus enantiomeric pairs.

5.5   Bridged Bicyclic Hydrocarbons

In the case of bridged bicyclic hydrocarbons, the isomerism also occurring there, which is quite similar to the *cis-/trans-*isomerism, must be replaced by the term *endo-/exo-*isomerism in order to be able to make clear statements about the spatial shape of the molecule.

Starting from the two atoms forming the bridgehead, the length of each bridge is determined first. The longest bridge is then the "first," the shortest the "third" bridge. The *endo*-isomer (Greek *endo* = inside) is now the isomer where the shortest bridge and the atoms with the lowest priority lie in relation to each other. In the *exo*-isomer (Greek *exo* = outside), these atoms lie away from each other.

If a hydrogen atom is substituted at the shortest bridge, then the *syn/anti* notation comes into play. *syn* (Greek: *syn* = together) is the position where the substituents face each other, and *anti* (Greek: *anti* = opposite) is the position where the substituent of the shortest bridge faces away from the other substituent.

Both *endo-/exo-* and *syn/anti* isomers are diastereomeric to each other.

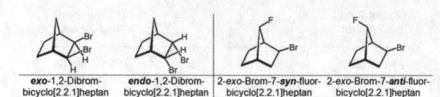

| *exo*-1,2-Dibrom-bicyclo[2.2.1]heptan | *endo*-1,2-Dibrom-bicyclo[2.2.1]heptan | 2-exo-Brom-7-*syn*-fluor-bicyclo[2.2.1]heptan | 2-exo-Brom-7-*anti*-fluor-bicyclo[2.2.1]heptan |

Configuration Isomers: Optical Isomerism

6

Configuration means the spatial arrangement of the atoms of a molecule, i.e., its spatial (structure) construction—but without taking into account possible rotations around single bonds.

Configuration isomers, which are caused by different spatial arrangement of the substituents on double bonds or on ring systems, have already been discussed in the section "geometric isomers" (Chap. 5). These compounds are usually achiral. Now we wil have a look at the chiral configuration isomers.

6.1 Chirality

If a C-atom is surrounded by four different substituents, there are two different, noncongruent but mirror image forms due to the tetrahedral arrangement around the C-atom. These mirror image forms are called enantiomers. Pairs of enantiomers only differ from each other in their optical activity and in their biological effect (see Sect. 4.1).

The in this way substituted C-atom is called the stereo center or stereogenic center (formerly also asymmetric C-atom). The enantiomers are named according to the CIP rules (see Sect. 5.3). Here each individual stereo center is designated by (R) or (S). An older form of nomenclature, the D- /L- nomenclature, is still used today for sugars and partly for amino acids.

Chirality does not only occur in compounds with stereo centers (so-called central chirality). Axial chirality occurs, for example, with sterically hindered biphenyls or with spiro compounds. Planar chirality can for example occur with (E)-cyclooctene. Helical chirality is found with helical structures.

© Springer Fachmedien Wiesbaden GmbH, part of Springer Nature 2021
T. Schmiermund, *Introduction to Stereochemistry,* Springer essentials,
https://doi.org/10.1007/978-3-658-32035-5_6

6.2   (*R*)/(*S*) Nomenclature

In the case of optically active compounds, i.e., enantiomers and diastereomers, the absolute configuration (= *R/S* nomenclature) is specified in such a way that the priorities of the substituents on *each* individual chiral C-atom are determined first (CIP rules, see Sect. 5.3). The molecule is then rotated so that the C-atom with the lowest priority points backward (= away from observer). The substituents now visible to the observer form the triangular base of the tetrahedron under consideration.

If the priority of the substituents $1 \rightarrow 2 \rightarrow 3$ runs to the right (= clockwise), then this C-atom holds *R* configuration (lat. *rectus* = right). If the substituent moves to the left (= counterclockwise), it holds *S* configuration (lat. sinister = left) (see Fig. 6.1).

The indication *R* or *S* must be given for each individual chiral C-atom.

6.2.1   (*l*)-/(*u*)-Nomenclature

If there are two stereo centers in a molecule, the two molecules with (R,R) or (S,S) configuration are also called (*l*) configuration (*like* = equal), and the two molecules with (R,S) or (S,R) configuration are called (*u*) compounds (*unlike* = unequal).

For example, (*l*)-tartaric acid is thus the pair (2R,3R)-tartaric acid (D-tartaric acid) and (2S,3S)-tartaric acid (L-tartaric acid). The (*u*)-tartaric acid is (2S,3R)-tartaric acid (meso-tartaric acid). See also Fig. 4.1.

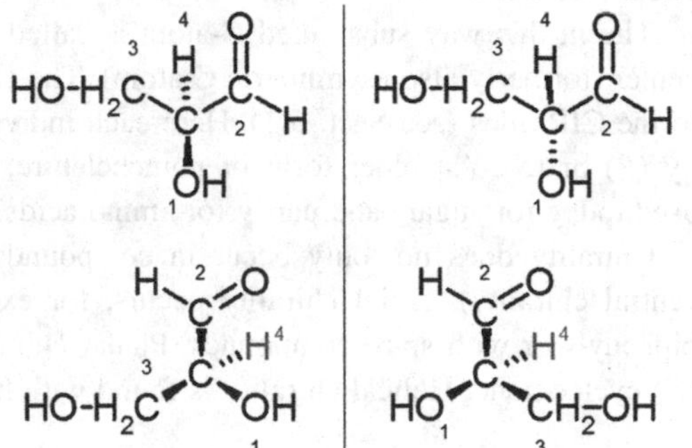

Fig. 6.1 (*R*)/(*S*) nomenclature using glyceraldehyde as an example, indicating the priority of the substituents

(S)-Glyceraldehyde (↺) | (R)-Glyceraldehyde (↻)

6.3   D/L **configuration, Fischer projection**

Before it was possible to precisely determine the stereochemical structure of molecules using spectroscopic methods, the structure and spatial organization of molecules had to be elucidated by means of a wide variety of chemical reaction sequences. For this purpose, a molecule was decomposed step by step. From the fragments, the overall structure of the starting compound was then deduced. For chiral compounds, glyceraldehyde was chosen as the standard. The right-turning (+) isomer was designated D (lat. *dextro* = right), and the left-turning (−) isomer L (lat. *laevo* = left).

Subsequently, any compound that could be related to D-(+)-glyceraldehyde by a series of degradation reactions and transformations was given the designation D and any compound to be related to L-(−)-glyceraldehyde was given the designation L.

To map three-dimensional structures of linear chiral compounds, the so-called Fischer projection was used. This projection was often used for molecules with several adjacent stereo centers. Even if this form of representation is no longer used today: On the naming of sugars and amino acids the Fischer projection was used in older books (see Fig. A.3).

Rules

- The chain of C-atoms is drawn from top to bottom (never horizontally). Here, the C-atom with the highest oxidation number comes on top and is numbered "1."
- Horizontal lines point out of the plane toward the observer.
- Vertical lines run backwards away from the observer.
- The C-atoms of the stereo centers are not written.
- Sugar: The position of the substituent (usually –OH) of the lowest stereo center indicates the configuration: D if the substituent is on the right, and L if it is on the left.
- Amino acids: Depending on whether the amino group (–NH₂) is shown on the right or left side, the D- or L-configuration results.

As it is not possible to use the D-, L-nomenclature in connection with the Fischer projection to give a separate configuration for all stereo centers, it is/was unavoidable to give different names to all diastereomers. Figure A.3 shows this as an example for the family of D-aldoses.

6.3.1   Comments

Please note that the nomenclatures according to $(R)/(S)$ and D/L are due to different reasons: The $(R)/(S)$ nomenclature is based on fixed rules, the D-/L-nomenclature is based on experimental results. (R) or (S) does not indicate whether a substance belongs to the D- or L-series or whether the optical activity is (+) or (−).

- Never try to classify a molecule as D or L based of of its structure.
- Never try to predict the optical rotation (+ or −) based on the structure.

6.3.2   Convert Fischer Projection to Wedge Formula

Several steps are necessary to develop the wedge formula from the Fischer projection. This will be illustrated using the example of D-glucose (dextrose, systematically: (2R,3S,4R,5R)-pentahydroxyhexanale)

a) Draw the Fischer projection
b) Mark the horizontal bindings as pointing outwards the plane.
c) Invert the bonds at every second stereogenic center (here: the 3rd and 5th C-atom). Note that this causes a lateral inversion of the OH groups.
d) Turn the molecule a 90° angle to the right. Draw the carbon chain so that the odd numbered C-atoms are low and the even numbered C-atoms are high. All OH groups in c) on the right side are pointing toward the observer, and those on the left are pointing away from the observer.

a Fischer Projection   **b** Meaning of the Fischer projection   **c** Inversion   **d** Rotation by 90°

6.4   *Threo/erythro* Isomers

The prefixes *threo-* and *erythro-* are derived from D-threose and D-erythrose, respectively, and were used for substances with exactly two directly adjacent stereo centers. If the residues at these stereo centers (as seen in the Fischer projection) are alternately arranged, the *threo* isomer is involved. If arranged on the same side, it is the *erythro* compound (compare Figs. 4.2 and A.3).

These designations are no longer recommended today. Instead, the $(R)/(S)$ or the $(l)-/(u)$-nomenclature should be used.

6.5   Axial chirality

Biphenyls are freely rotatable around the axis connecting the two rings. By introducing substituents in the o,o' position (1,6 position) on both rings, the free rotation can be massively restricted, thus creating axial chirality. If the energy barrier is very high, both forms can be isolated and are no longer regarded as conformers. The first compound in which axial chirality was discovered is 6,6'-dinitrodiphenoic acid (6,6'-dinitro-biphenyl-2,2'-dicarboxylic acid).

(aS)-   (aR)-
6,6'-Dinitro-biphenyl-2,2'-dicarboxylic acid

Spiro compounds also form chiral molecules. The two rings are connected by a tetrahedrally coordinated carbon atom and are therefore perpendicular to each other. Therefore, even compounds that look symmetrical at first glance are surprisingly chiral. In order to indicate axial chirality (without a stereo center), the (R) or (S) to indicate the configuration is preceded by an a (for axial).

(aS)-	**(aR)-**	**(aS)-**	**(aR)-**
Spiro[4.4]nonadiene		**Spiro[5.5]undecane-1,7-dione**	

6.6   Planar Chirality

(*E*)-cyclooctene is to serve as an example of planar chirality. This compound has no axis of rotation, so that two enantiomers exist: (p*R*)-(*E*)-cyclooctene and (p*S*)-(*E*)-cyclooctene. The prefix p indicates planar chirality.

(p*R*)-	**(p*S*)-**
(*E*)-Cyclo-octene	

6.7   Helical Chirality

Helicenes are probably the simplest helical compounds. They are helical compounds made up of ortho-annealed aromatic rings.

If the course of the screw (seen as away from the observer, so to say "downward") follows the clockwise direction ("right-hand thread"), the compound has (*P*) configuration (*P* = plus). If the screw moves counterclockwise ("left-hand thread"), the (*M*) configuration (*M* = minus) is present. This special type of axial chirality is also called helicity.

The two enantiomers of hexahelicene (also: [6]helicene or phenanthro[3,4-c]phenanthrene) are shown as an example:

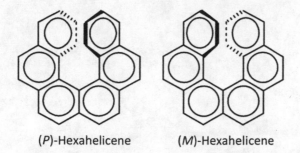

(*P*)-Hexahelicene   (*M*)-Hexahelicene

Configuration Isomers: Sugar

7

Sugars (monosaccharides), together with oligosaccharides (consisting of 2 to 6 sugar molecules) and polysaccharides (consisting of up to several 1000 sugar molecules, e.g., cellulose) belong to the so-called carbohydrates. The name is derived from the general molecular formula $C_nH_{2n}O_n$, which can also be understood as $C_n(H2O)_n$.

Monosaccharides are polyvalent alcohols in which one OH-group has been oxidized to carbonyl ($-C(O)-$). If the carbonyl group is located on the C1 atom, it is an aldose (of aldehyde; $-CHO$). If it is on the C2 atom, the sugar is called ketose (from ketone, $-C(O)-$).

Depending on the number of C-atoms, monosaccharides are called triose, tetrose, pentose, hexose, etc. Accordingly, the two simplest sugars are the aldotriose glyceraldehyde ($HOCH_2-CH(OH)-CHO$) and the ketotriose dihydroxyacetone ($HOCH_2-C(O)-CH_2OH$). Note that two of these C-atoms (the two ends of the chain) are always nonchiral and that these must be deducted to calculate the maximum possible number of diastereomers.

Often monosaccharides are imaged in the Fischer projection, because the D-, L-configuration can be represented well with this method. In reality, the monosaccharides are present as cyclic hemiacetal forms due to oxo-cyclo tautomerism. Furanoses (cyclo-[C_5O], 6-ring) and pyranoses (cyclo-[C_4O], 5-ring) are distinguished depending on whether a tetrahydrofuran or tetrahydropyran ring is formed. The suffix -ose is accordingly replaced by -ofuranose or -opyranose.

This ring-shaped structure is sometimes reproduced by an extended Fischer projection. Here, the ring oxygen atom is either pulled to the right and placed in the middle of the existing vertical C-chain or (alternatively) this oxygen atom is left at its Fischer projection position and the O–C bond is pulled longer (upward).

© Springer Fachmedien Wiesbaden GmbH, part of Springer Nature 2021
T. Schmiermund, *Introduction to Stereochemistry,* Springer essentials,
https://doi.org/10.1007/978-3-658-32035-5_7

7.1　Anomers

The intramolecular ring closure to the hemiacetal form of the monosaccharides is associated with the formation of a new stereo center. This is called the anomeric C-atom. From one oxo-form, i.e., from one enantiomer, two chiral diastereomeric cyclohemiacetal forms can thus be formed, which in turn are called anomers (Greek *ano* = upper).

The anomers differ in the spatial orientation of the OH-group on the C1 atom. For monosaccharides of the D- series, the following applies: If the OH-group is axially oriented (= downward in the Haworth ring formula or to the right in the extended Fischer projection), this is called the α-form. If the orientation is equatorial (Haworth: top, Fischer: left), it is the β-form. The opposite is true for simple sugars of the L- series.

This hydroxyl group, which is important for the differentiation of the two anomers, is sometimes also called glycosidic OH-group or anomeric hydroxyl group.

7.2　Display Variants

Better than the (extended) Fischer projection, the structure of the cyclic hemiacetals can be represented by perspective representation in the chair form (without H-atoms).

Another possibility is the Haworth ring formula. In this formula, the oxygen atom belonging to the ring is always on the top right in the case of furanoses and on the top middle in the case of pyranoses. The ring is drawn planar. To convert the Fischer projection into the Haworth ring formula, remember FLOH: **F**ischer on the **l**eft is on top at **H**aworth (German: *Was bei Fischer links ist, ist oben bei Haworth*).

Furthermore, an illustration in the wedge formula is possible. Figure 7.1 shows the different possibilities using the example of D- glucose.

7.3　Nomenclature of Monosaccharides

The possibilities of representation are as manifold as the possibilities of naming these substances. This different nomenclature can be derived directly from the different representations. This will be demonstrated usingD- glucose as an example (see Fig. 7.1).

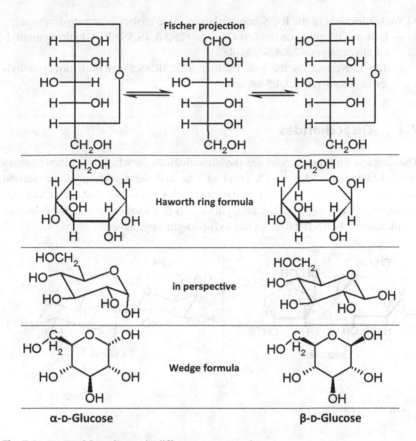

Fig. 7.1 α-D- and β-D-glucose in different representations

- Fischer projection in the D-, L- nomenclature:
  - Inner (right) OH-group on the C1 atom: α-D-glucose or α-D-glucopyranose
  - Outer (left) OH-group on the C1 atom: β-D-glucose or β-D-glucopyranose
- Haworth ring formula in the D-, L- nomenclature:
  - Bottom OH-group on C1 atom: α-D-glucopyranose
  - Top OH-group on C1 atom: β-D-glucopyranose
- Haworth ring formula in the R-, S-nomenclature:
  - Bottom OH-group on the C1 atom: (1S,2R,3R,4R,5R)-glucopyranose
  - Top OH-group on the C1 atom: (1R,2R,3R,4R,5R)-glucopyranose

- Wedge formula in the R-, S-nomenclature (base molecule: tetrahydropyran):
  - Bottom OH-group on the first C-atom: (2S,3R,4S,5S,6R)-6-hydroxymethyl-tetrahydro-pyran-2,3,4,5-tetraole
  - Top OH-group on the first C-atom: (2R,3R,4S,5S,6R)-6-hydroxymethyl-tetrahydro-pyran-2,3,4,5-tetraole

7.4   Disaccharides

The simplest oligosaccharides are the disaccharides, in which two simple sugars are linked together. As the C1 atom of one and the C4 atom of the second sugar are linked together, this is called a β(1,4)glucosidic linkage. Examples are sucrose ('cane sugar', α-D-glucopyranosyl-β-D-fructofuranoside) and β-lactose ('milk sugar', 4-O(β-D-galactopyranosyl)-β-D-glucopyranose).

Sucrose

β-Lactose

Conformation Isomers

8

Chemical bonds are not rigid and should therefore not be considered as if they were solid bars. Rather, bonds are subject to various movements whose strength ("deflection") is determined by the temperature of the substance, among other things. These are primarily:

- Translation movements
  - Valence or stretch oscillations (change in the length of a bond)
  - Deformation or bending vibrations (change of the angle of a bond)
- Rotational movements (rotation around an axis)

These oscillations are of greater interest in the context of IR spectroscopy (compare corresponding literature). Here we only consider the rotation of a molecule part around the "axis of rotation" of the single bond. A free rotation around double or triple bonds is not possible, because this would require π-bonds to be released and then reattached.

The angle by which a part of a molecule is rotated around the single bond in question is known as the dihedral or torsion angle.

8.1 Visualization possibilities

Sawhorse projection

- In the *sawhorse* projection, the σ bond represents the "support" of the sawhorse. The sawhorse feet are the rotating parts of the molecule.

© Springer Fachmedien Wiesbaden GmbH, part of Springer Nature 2021
T. Schmiermund, *Introduction to Stereochemistry,* Springer essentials,
https://doi.org/10.1007/978-3-658-32035-5_8

41

Wedge formula

- In the so-called wedge formula (also: perspective formula), the bonds shown as wedges point forward out of the paper plane (i.e., toward the viewer), and the bonds shown as dotted lines point backward (away from the viewer). The solid lines lie in the paper plane. The entire molecule is viewed from the side.

Newman projection

- Here the molecule is viewed from the front: From the C1 atom to the C2 atom. The solid lines are bonds to the front carbon atom (C1), which is located at the intersection of the lines. The lines ending at the circle are bonds to the rear C-atom (C2), which is covered by the front C-atom.

The dihedral angles are easiest to see in this representation.

8.2   Ethane Conformers

In ethane (C_2H_6), the two carbon atoms are connected by a freely rotatable (and rotationally symmetric) σ-bond. The rotation around this single bond of one of the two CH_3-groups results in different spatial arrangements, which differ in their energy content. The resulting spatial isomers are called conformers. In ethane, there are only two conformers: One in which all hydrogens are behind each other (ecliptic or hidden) and one in which the hydrogens are "on gap" (staggered). The *staggered* conformation has about 12 kJ mol^{-1} less energy than the *eclipsed* one. The dihedral angles (θ = theta) for the ecliptic conformations are 0°, 120°, 240° and 360°. For the staggered conformations, the corresponding angles are 60°, 180° and 300° (see Fig. 8.1).

8.3   Butane Conformers

If the conformers of propane (C_3H_8) do not differ from those of ethane (energy barrier approx. 14 kJ mol^{-1}), the situation is somewhat more difficult with butane (C_4H_{10}). Here, more than two conformers are possible, depending on the angle of rotation.

If we consider n-butane as 1,2-dimethyl-ethane, it does not differ from ethane in terms of its basic forms: The dihedral angles θ = 0°, 120° and 240° are ecliptic, the angles 60°, 180° and 300° are staggered.

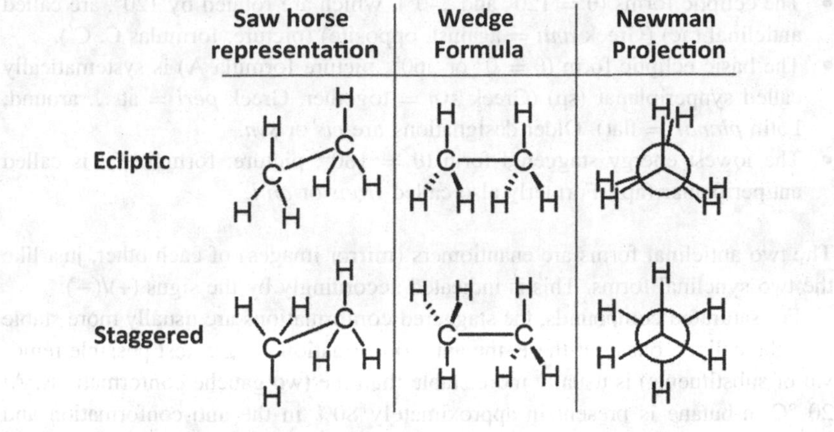

	Saw horse representation	Wedge Formula	Newman Projection
Ecliptic			
Staggered			

Fig. 8.1 Conformations of the ethane

However, the different dihedral angles are given their own designations to enable them to be distinguished from one another (see Fig. 8.2):

- The staggered forms, which are rotated 60° to the right or left ($\theta = 60°$ or $300°$) are also called gauche conformations (French *gauche* = skew). The systematic name is synclinal (sc) (Greek *syn* = together, with; Greek *klinein* = to lean against, incline) (picture, formulas B, B′).

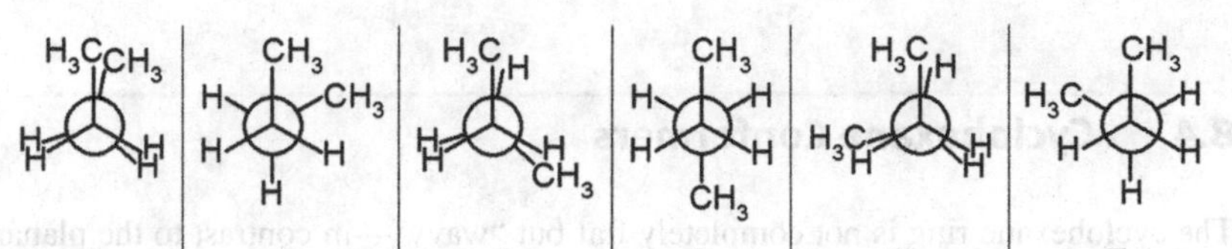

A	B	C	D	C′	B′
0°/360°	60°	120°	180°	240°	300°
+18.8 kJ mol⁻¹	+3.8 kJ mol⁻¹	+15.8 kJ mol⁻¹	0 kJ mol⁻¹	+15.8 kJ mol⁻¹	+3.8 kJ mol⁻¹
Ecliptic	Staggered	Ecliptic	Staggered	Ecliptic	Staggered
cis/syn synperiplanar ± sp	gauche (+)-synclinal + sc	– (+)-anticlinal + ac	trans/anti antiperiplanar ± ap	– (–)-anticlinal – ac	gauche (–)-synclinal – sc

Fig. 8.2 Conformations of butane; with indication of the dihedral angle, the relative energy difference, the form and the various designations

- The ecliptic forms (θ = 120° and 240°), which are rotated by 120°, are called anticlinal (ac) (Greek *anti* = against, opposite) (picture, formulas C, C′).
- The basic ecliptic form (θ = 0° or 360°, picture formula A) is systematically called synperiplanar (sp) (Greek *syn* = together, Greek *peri* = at ... around; Latin *planar* = flat). Older designations are *cis* or *syn*.
- The lowest energy staggered form (θ = 180°, picture, formula D) is called antiperiplanar (ap). Formerly also called *trans* or *anti*.

The two anticlinal forms are enantiomers (mirror images) of each other, just like the two synclinal forms. This is indicated accordingly by the signs (+)/(−).

For saturated compounds, the staggered conformations are usually more stable than the ecliptic ones. Of these, the anti-conformation (= greatest possible removal of substituents) is usually more stable than the two gauche conformations. At 20 °C, n-butane is present in approximately 80% in the anti-conformation and approximately 20% in the two gauche forms.

On the other hand, for compounds with double bonds (e.g., alkenes, aldehydes, ketones), the conformation in which the double bond is arranged ecliptically is preferred. If hydrogen bonds can be formed, the gauche form is often stabilized.

Acetaldehyde
H₃C-CHO

2-Chloro-ethanol
ClH₂C-CH₂OH

8.4   Cyclohexane Conformers

The cyclohexane ring is not completely flat but "wavy"—in contrast to the planar benzene. There are two basic conformations: A *chair* form and a *boat* form. The chair conformation is the more stable one.

The rotation of all C-atoms around their single bond results in a so-called ring inversion. In simplified terms, the chair form (I) becomes the boat form, which can then turn over into the chair form (II) (for details, see textbooks of organic chemistry). There are therefore two equivalent chair forms (see Sect. 5.2).

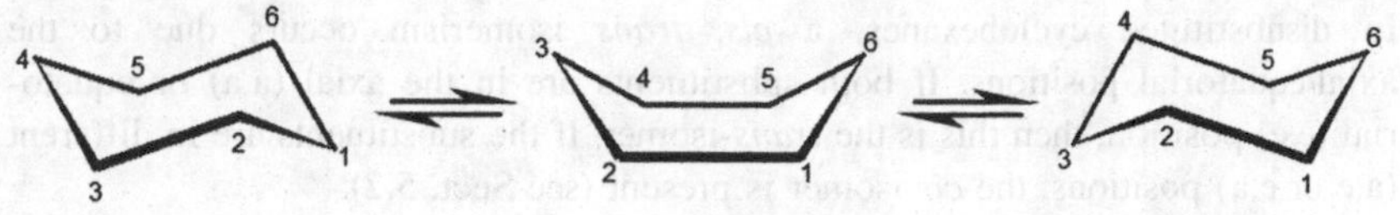

Chair form (I)   Boat form   Chair form (II)

Due to the tetrahedral alignment of the C-bonds, the H-atoms or substituents are in two different positions relative to the central ring plane: Six bonds are aligned perpendicular to the ring plane, oriented alternately upward and downward and are referred to as axial (a). Six others are arranged at about 70° to the main axis and are also oriented alternately up and down. They are called *equatorial* bonds (e).

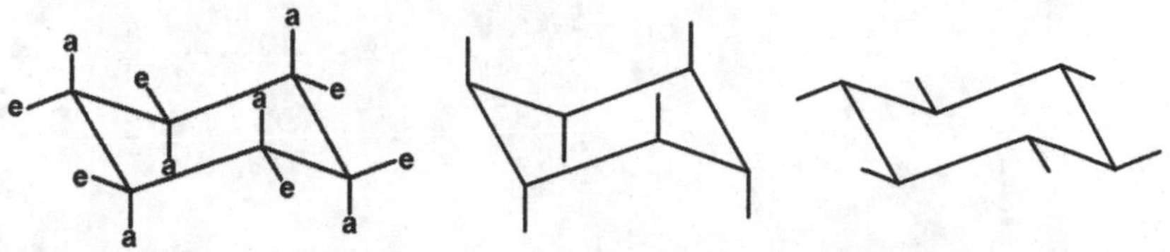

axial (a) and equatorial (e) bonds

axial bonds only

equatorial bonds only

The ring inversion now causes all axial bonds to move to equatorial position— and vice versa. Monosubstituted cyclohexanes therefore have two conformers: one with an axial and one with an equatorial substituent.

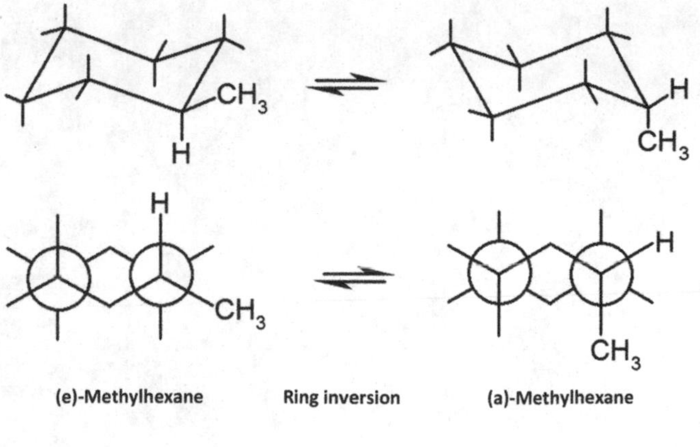

(e)-Methylhexane   Ring inversion   (a)-Methylhexane

In disubstituted cyclohexanes, a *cis*, *trans* isomerism occurs due to the axial/equatorial positions. If both substituents are in the axial (a,a) or equatorial (e,e) position, then this is the *trans*-isomer. If the substituents are in different (a,e or e,a) positions, the *cis*-isomer is present (see Sect. 5.2).

What You Learned From This *essential*

- Nomenclature and variants of the stereochemistry of organic compounds
- Common and less common terms in stereochemistry
- Different constitutional isomers, differences and similarities
- Enantiomers and diastereomers: connections and differentiation
- Nomenclature rules according to Cahn, Ingold and Prelog (CIP-Convention)
- Different display variants for sugar molecules

© Springer Fachmedien Wiesbaden GmbH, part of Springer Nature 2021
T. Schmiermund, *Introduction to Stereochemistry,* Springer essentials,
https://doi.org/10.1007/978-3-658-32035-5

Overviews

(See Figs. A.1, A.2 and A.3).

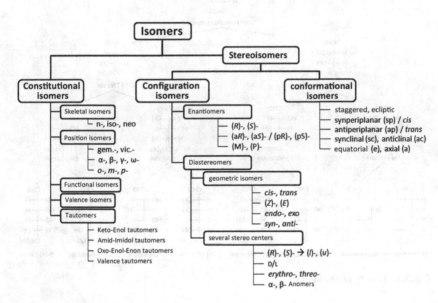

Fig. A.1 Overview isomers

© Springer Fachmedien Wiesbaden GmbH, part of Springer Nature 2021
T. Schmiermund, *Introduction to Stereochemistry,* Springer essentials,
https://doi.org/10.1007/978-3-658-32035-5

Nature of the isomers	Common features	Differences	Physical properties	Chemical properties	Descriptor/Note	Chap.
Functional isomers	Molecular formula	Funct. groups	All physical properties of the isomers differ	Different reactivity	-	3.3
Skeletal isomers		C-skeleton			n-, iso-, neo-	3.1
Position isomers	Skeleton, Functions	Position on the C-skeleton			α-, β-, γ- o-, m-, p-	3.2
Valence isomers		Bonds			-	3.4
Tautomers		Functions			Keto-Enol-T.	3.5
Stereoisomers Diastereomers - geometrical Isomers	Constitution	Relative arrangement on double bond or ring		Transformation of isomers only possible by breaking an rebonding of bonds	cis-, trans- (Z)-, (E)- endo-, exo- syn-, anti-	5.1/5.2 5.4 5.5 5.5
- several centers		Relative arrangement of chiral groups			(R)-, (S)- D-, L- erythro-, threo- α-, β-Anomers	6.3 6.4 6.5 7.1
Enantiomers		Chiral molecules, image/mirror image	Different behavior against polarized light	Differences only in chiral reactants	(R)-, (S)- (M)-, (P)-	6.6/6.7 6.8
Conformational isomers		Different torsion angles	Isomers are usually not separable	Isomerization without bond b reakage	ecliptic, staggered gauche	8

Fig. A.2 Properties of different isomers

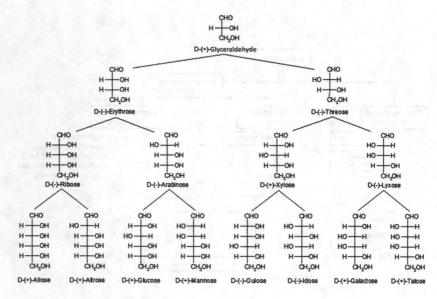

Fig. A.3 Family of D-aldoses

References

Allinger, N. L., Cava, M. P., de Jong, D. C., Johnson, C. R., Lebel, N. A., & Stevens, C. L. (1980). *Organische Chemie* . Berlin: De Gruyter (original edition: Organic Chemistry (1976), New York).

Beyer, H., & Walter, W. (1988). *Lehrbuch der Organischen Chemie* (21st ed.). Stuttgart: Hirzel.

Bruice, P. Y. (2007). *Organische Chemie* (5. Aufl.). München: Pearson-Studium (original edition: Organic Chemistry (2007), Pearson Ediucation).

Christen, H. R. (1982). *Grundlagen der organischen Chemie* (5th ed.). Frankfurt a. M.: Otto Salle.

Clayden, J., Greeves, N., & Warren, S. (2013). *Organische Chemie* (2. Aufl.). Heidelberg: Springer (original edition: Organic Chemistry (2012) Oxford University Press).

Dale, J. (1978). Stereochemie und Konformationsanalyse. Weinheim: Verlag Chemie.

Dickerson, R. E., Gray, H. B., & Haight, G. P. (1978).*Prinzipien der Chemie*. Berlin: De Gruyter (original edition: Chemical Principles, Second Edition (1974) W. A. Benjamin Inc.

Felixberger, J. K. (2017). *Chemie für Einsteiger*. Heidelberg: Springer.

Hauptmann, S. (1988). *Organische Chemie*. Berlin: VEB Deutscher Verlag für Grundstoffindustrie.

Hellwich, K. H. (1998). *Chemische Nomenklatur* (3rd ed.). Eschborn: Govi.

Hellwich, K. H. (2007). *Stereochemie: Grundbegriffe* (2nd ed.). Heidelberg: Springer.

Hellwinkel, D. (2005). *Die systematische Nomenklatur der organischen Chemie* (5th ed.). Heidelberg: Springer.

Latscha, H. P., & Klein, H. A. (1997). *Organische Chemie* (4th ed.). Heidelberg: Springer.

Morrison, R. T., & Boyd, R. N. (1986). *Lehrbuch der Organischen Chemie* (3. Aufl.). Weinheim: Verlag Chemie (original edition: Organic Chemistry, 4th edition (1983) Boston: Allyn and Bacon).

Mortimer, C. E. (1996). *Chemie* (6. Aufl.). Stuttgart: Georg Thieme (original edition: Chemistry (1993) Belmont: Wadsworth Publishing).

Neubauer, D. (2014). *Kekulés Träume*. Heidelberg: Springer.

Schmiermund, T. (2019). *Das Chemiewissen für die Feuerwehr*. Heidelberg: Springer.

Streitwieser, A., Heathcock, C. H., & Kosower, E. M. (1994). *Organische Chemie* (2. Aufl.). Weinheim: Wiley-VCH (original edition: Introduction to Organic Chemistry 4th edition (1992) Macmillian Publishing Company).

© Springer Fachmedien Wiesbaden GmbH, part of Springer Nature 2021   51
T. Schmiermund, *Introduction to Stereochemistry,* Springer essentials,
https://doi.org/10.1007/978-3-658-32035-5

Vollhardt, K. P. C., & Schore, N. E. (1995). *Organische Chemie* (2. Aufl.). Weinheim: Wiley-VCH (original edition: Organic Chemistry 2nd edition (1994) New York: W. H. Freeman and Company).

Wawra, E., Dolznig, H., & Müllner, E. (2010). *Chemie erleben* (2nd ed.). Wien: Facultas.

Printed in the United States
by Baker & Taylor Publisher Services